U.S. ENVIRONMENTAL PROTECTION AGENCY

OFFICE OF INSPECTOR GENERAL

Unless California Air Resources Board Fully Complies With Laws and Regulations, Emission Reductions and Human Health Benefits Are Unknown

Report No. 14-R-0130   March 6, 2014

Report Contributors:

Darren Schorer
Lela Wong

Abbreviations

CARB	California Air Resources Board
CFR	Code of Federal Regulations
DERA	Diesel Emissions Reduction Act
EPA	U.S. Environmental Protection Agency
FY	Fiscal Year
NOx	Nitrogen Oxide
NREC	National Railway Equipment Company
OIG	Office of Inspector General
OMB	Office of Management and Budget

Cover photo: An example of a BNSF Railway Company switch-yard locomotive located at the Hobart rail yard in Commerce, California. (EPA photo)

Hotline

To report fraud, waste or abuse, contact us through one of the following methods:

email:	OIG_Hotline@epa.gov
phone:	1-888-546-8740
fax:	1-202-566-2599
online:	http://www.epa.gov/oig/hotline.htm
write:	EPA Inspector General Hotline 1200 Pennsylvania Avenue, NW Mailcode 2431T Washington, DC 20460

Suggestions for Audits or Evaluations

To make suggestions for audits or evaluations, contact us through one of the following methods:

email:	OIG_WEBCOMMENTS@epa.gov
phone:	1-202-566-2391
fax:	1-202-566-2599
online:	http://www.epa.gov/oig/contact.html#Full_Info
write:	EPA Inspector General 1200 Pennsylvania Avenue, NW Mailcode 2410T Washington, DC 20460

At a Glance

Why We Did This Review

The U.S. Environmental Protection Agency (EPA) Office of Inspector General (OIG) conducted this examination to determine whether the costs claimed under cooperative agreement 2A-00T13801 awarded under the American Recovery and Reinvestment Act of 2009 to the California Air Resources Board (CARB) were reasonable, allowable and allocable in accordance with applicable laws, regulations, and cooperative agreement terms and conditions. The OIG also sought to determine whether cooperative agreement objectives were met.

This report addresses the following EPA themes:

- *Addressing climate change and improving air quality.*
- *Making a visible difference in communities across the country.*

For further information, contact our public affairs office at (202) 566-2391.

The full report is at:
www.epa.gov/oig/reports/2014/20140306-14-R-0130.pdf

Unless California Air Resources Board Fully Complies With Laws and Regulations, Emission Reductions and Human Health Benefits Are Unknown

What We Found

Our examination disclosed material weaknesses in CARB's compliance with laws, regulations, and the terms and conditions of the cooperative agreement. Specifically, CARB did not comply with the requirement of the cooperative agreement and the Energy Policy Act of 2005 to scrap or remanufacture the old engines. CARB also did not accurately report jobs created or retained or provide actual emissions reduction calculations, as required under the cooperative agreement. In addition, CARB paid contract costs that were not in accordance with contract terms.

> CARB did not fully comply with the Energy Act and cooperative agreement requirements, resulting in potential overpayment of $8,866,000.

CARB completed the locomotive repower according to the workplan. However, CARB has not demonstrated that it met the cooperative agreement objective for achieving significant emissions reduction as CARB did not provide actual emissions benefit calculations.

Recommendations and Corrective Actions

For the contract terms issue, we recommended that the Region 9 Regional Administrator disallow and recover ineligible costs of $94,109 claimed under the cooperative agreement and require CARB to establish internal controls prior to any future awards. In response to draft report, CARB repaid the $94,109 to the EPA. CARB's contractor—BNSF Railway Company—also signed an agreement on November 18, 2013, to scrap or remanufacture the replaced engines. We recommend that the region verify that CARB and BNSF comply with the agreement and document the scrap or remanufacture. The region should recover the federal share of $8,771,891 claimed in the event CARB violates the November 18, 2013, agreement.

We also made recommendations related to jobs reported as created or retained, developing a more accurate calculation of project results based on actual fuel usage, and adjusting the reporting of CARB's project results. In response to the draft report, CARB submitted the job report corrections. Region 9 and CARB disagreed with the two recommendations relating to project results.

Noteworthy Achievements

CARB and the BNSF Railway Company repowered 11 locomotives instead of the eight required by the cooperative agreement because of cost reductions achieved by the manufacturer and a higher contractor contribution percentage than what was required under the agreement. The subcontractor also accomplished construction of the 11 locomotives in less than 4 months in order to meet the Recovery Act deadline of September 30, 2010.

UNITED STATES ENVIRONMENTAL PROTECTION AGENCY
WASHINGTON, D.C. 20460

March 6, 2014

MEMORANDUM

SUBJECT: Unless California Air Resources Board Fully Complies with Laws and Regulations, Emission Reductions and Human Health Benefits Are Unknown
Report No. 14-R-0130

FROM: Arthur A. Elkins Jr.

TO: Jared Blumenfeld, Regional Administrator
Region 9

This is our report on the subject examination conducted by the Office of Inspector General (OIG) of the U.S. Environmental Protection Agency (EPA). This report contains findings that describe problems the OIG has identified and corrective actions the OIG recommends. This report represents the opinion of the OIG and does not necessarily represent the final EPA position. In accordance with established audit-resolution procedures, EPA managers will make final determinations on matters in this report.

Action Required

In accordance with EPA Manual 2750, you are required to provide us your proposed management decision on the findings and recommendations contained in this report before you formally complete resolution with the recipient. Your proposed management decision is due in 120 days, or on July 7, 2014. To expedite the resolution process, please email an electronic version of your proposed management decision to adachi.robert@epa.gov.

Your response will be posted on the OIG's public website, along with our memorandum commenting on your response. Your response should be provided as an Adobe PDF file that complies with the accessibility requirements of Section 508 or the Rehabilitation Act of 1973, as amended. The final response should not contain data that you do not want to be released to the public; if your response contains such data, you should identify the data for redaction or removal along with corresponding justification. We will post this report to our website at http://www.epa.gov/oig.

If you or your staff have any questions regarding this report, please contact Rich Eyermann, acting Assistant Inspector General for Audit, at (202) 566-0565 or eyermann.richard@epa.gov; or Robert Adachi, Product Line Director, at (415) 947-4537 or adachi.robert@epa.gov.

Unless California Air Resources Board
Fully Complies With Laws and Regulations,
Emission Reductions and Human Health Benefits Are Unknown

14-R-0130

Table of Contents

Chapters

Appendices

Chapter 1
Independent Accountant's Report

As part of our oversight of assistance agreement awards made by the U.S. Environmental Protection Agency (EPA), the EPA's Office of Inspector General (OIG) examined the costs claimed under Cooperative Agreement No. 2A-00T13801 awarded to the California Air Resources Board (CARB). The OIG conducted the examination to determine whether the costs claimed under the cooperative agreement were reasonable, allowable and allocable in accordance with the Code of Federal Regulations (CFR) under 40 CFR Part 31, *Uniform Administrative Requirements for Grants and Cooperative Agreements to State and Local Governments*; 2 CFR Part 225, *Cost Principles for State, Local, and Indian Tribal Governments* (Office of Management and Budget Circular A-87); and the terms and conditions of the cooperative agreement. We also reviewed CARB's accomplishment of the cooperative agreement's objectives and compliance with the following American Recovery and Reinvestment Act of 2009 requirements:

- Buy American requirement under Section 1605.
- Davis-Bacon Act wage requirements under Section 1606.
- Job reporting requirements under Section 1512.

By accepting the funding provided through the cooperative agreement, CARB is responsible for complying with these requirements. Our responsibility is to express an opinion on CARB's compliance based on our examination.

We conducted our examination in accordance with generally accepted government auditing standards issued by the Comptroller General of the United States and the attestation standards established by the American Institute of Certified Public Accountants. We examined, on a test basis, evidence supporting management's assertions and performed such other procedures as we considered necessary in the circumstances. We believe that our examination provides a reasonable basis for our opinion.

We conducted our audit work between March 11, 2013, and September 25, 2013. We contacted representatives from the EPA's Region 9 to gather information on criteria relevant to the cooperative agreement, obtain an understanding of the proposed project, and gather information concerning CARB's performance. Specifically, we performed the following steps:

- Reviewed the EPA's project and cooperative agreement files.
- Reviewed the application associated with the award.
- Reviewed award documents, including amendments.
- Reviewed CARB's original and revised work plans.
- Conducted interviews with EPA Region 9 staff.

On March 11, 2013, we made a site visit to CARB's office in Sacramento, California, to conduct interviews and obtain documentation needed to address our objectives. We performed the following steps:

- Reviewed recipient cooperative agreement files and interviewed recipient personnel to gain an understanding of the accounting system, internal controls, and the costs reported and work performed under the cooperative agreement.
- Reviewed CARB's request for proposal, bids submitted, contracts and subcontracts.
- Reviewed costs claimed under the cooperative agreement to determine whether the costs met applicable requirements under 40 CFR Part 31, 2 CFR Part 225, and the terms and conditions of the cooperative agreement.
- Reviewed CARB's compliance with Recovery Act requirements.
- Conducted a site visit to the BNSF Railway Company's rail yards to verify the existence of locomotives and completion of work in accordance with the work plan.
- Reviewed supporting documentation for reported emissions reductions.

CARB is responsible for establishing and maintaining effective internal control over compliance with the requirements of 40 CFR Part 31, 2 CFR Part 225, the Recovery Act, and the terms and conditions of the cooperative agreement. In planning and performing our examination, we considered CARB's internal control over compliance with the requirements listed above as a basis for designing our examination procedures for the purpose of expressing our opinion on compliance, but not for the purpose of expressing an opinion on the effectiveness of internal control over compliance. Accordingly, we do not express an opinion on the effectiveness of CARB's internal control over compliance.

Our consideration of internal control over compliance was for the limited purpose described in the preceding paragraph and was not designed to identify all deficiencies in internal control over compliance that might be significant deficiencies or material weaknesses; therefore, there can be no assurance that all deficiencies, significant deficiencies or material weaknesses have been identified.

A significant deficiency is a deficiency in internal control, or combination of deficiencies, that adversely affects that entity's ability to initiate, authorize, record, process or report data reliably in accordance with the applicable criteria or framework, such that there is more than a remote likelihood that a misstatement of the subject matter that is more than inconsequential will not be prevented or detected. A material weakness is a significant deficiency, or combination of significant deficiencies, that results in more than a remote likelihood that material misstatement of the subject matter will not be prevented or detected.

Our examination disclosed material weaknesses in CARB's compliance with laws, regulations and cooperative agreement requirements. In particular, CARB:

- Did not comply with the requirement of the cooperative agreement and the Energy Policy Act to scrap or remanufacture the old engines.
- Did not accurately report jobs created or retained.
- Did not provide actual emissions reduction calculations.

Our examination also disclosed a significant deficiency in internal controls that resulted in CARB paying contract costs that were not in accordance with contract terms.

As a result of the issues noted above, we questioned the entire $8,866,000 claimed under the cooperative agreement. In our opinion, CARB has not complied with the terms and conditions of cooperative agreement 2A-00T13801.

CARB completed the locomotive repower according to the workplan. Specifically, CARB repowered 11 locomotives (three more than the eight required under the cooperative agreement) using EPA-verified technology, and the repowered locomotives were operating in the required area. However, CARB has not demonstrated that it met the cooperative agreement objective for achieving significant emissions reduction as CARB did not provide actual emission benefit calculations.

Robert K. Adachi
Director, Forensic Audits
March 6, 2014

Chapter 2
Introduction

Purpose

The EPA OIG conducted this examination to determine whether the costs claimed under cooperative agreement 2A-00T13801 were reasonable, allocable and allowable in accordance with applicable federal requirements and the terms and conditions of the cooperative agreement. The OIG also reviewed CARB's compliance with select Recovery Act requirements and the accomplishment of cooperative agreement objectives.

Background

The Diesel Emissions Reduction Act (DERA) was signed into law in August 2005 under Title VII, Subtitle G, of the Energy Policy Act of 2005. DERA authorized $200 million per year from fiscal years (FYs) 2007 through 2011 (a total of $1 billion). The money enabled the EPA to fund programs to achieve significant reductions in diesel emissions (e.g., tons of pollution produced and diesel-emission exposures, particularly from fleets operating in areas designated by the agency as poor air quality areas). Of the authorized DERA amount, 70 percent is authorized for competitive national grants and low-cost revolving loans, as determined by the EPA Administrator. The remaining 30 percent is for state grants and loan programs. Congress appropriated a total of $219.1 million for the EPA under DERA for FYs 2008 through 2011. Congress appropriated an additional $300 million to the EPA in FY 2009 for DERA grants under the Recovery Act.

On July 10, 2009, the EPA used Recovery Act funds to award cooperative agreement 2A-00T13801 to CARB under the National Clean Diesel Funding Assistance Program. This award authorized federal funds of $8,888,888 to repower eight existing switch-yard locomotives with new Tier 3 nonroad engines. Total project costs were $12,000,000, which included the authorized federal funds and the recipient's share of $3,111,112. The project period was June 15, 2009, through December 31, 2010.

CARB proposed operating the repowered locomotives within rail yards located in the South Coast Air Basin. The EPA designated the South Coast Air Basin as a particulate matter nonattainment area. CARB conducted health risk assessments of 18 major rail yards and determined that 90 percent or more of switch-yard locomotive emissions are generated in and around large rail yards in California. CARB also determined that switch-yard locomotives can represent at least half of the locomotive diesel particulate matter within and around rail yards.

CARB estimates that a new locomotive engine could potentially provide up to a 90-percent reduction in nitrogen oxide (NOx) and particulate matter emissions compared to an older engine. Specifically, a repowered locomotive could provide reductions of up to 0.045 tons per day for NOx emissions and 0.0018 tons per day for particulate matter emissions. A 90-percent reduction could provide substantial cancer and noncancer health benefits to communities in the South Coast Air Basin, which consists of parts of Los Angeles and San Bernardino counties and all of Orange County (see figure 1 below).

Figure 1: Map of the South Coast Air Basin

Source: CARB website at www.arb.ca.gov/msprog/onroad/porttruck/maps/scab7map.pdf.

Noteworthy Achievements

CARB and BNSF repowered three locomotives in addition to the eight required by the cooperative agreement. In addition, the federal share of the project cost was $22,888 less than the authorized funding of $8,888,888. The additional repowered locomotives and the reduction in the federal share was because of cost reductions achieved by the manufacturer and because BNSF contributed funding in excess of the recipient's share of 25.93 percent required by the cooperative agreement. The subcontractor also accomplished construction of the 11 locomotives in less than 4 months in order to meet the Recovery Act deadline of September 30, 2010.

Chapter 3
Costs Claimed Were Not in Accordance With Contract Provisions

Our examination disclosed that CARB paid contractor costs that were not in compliance with contract provisions because CARB did not monitor contractor billings to ensure compliance. Specifically, the contractor billed costs in excess of CARB's share of the total costs, which included a remote-control system that was not for the locomotives under the cooperative agreement. As a result, we questioned $94,109 of the $8,866,000 claimed under the cooperative agreement.

Costs Billed by the Contractor Exceeded Contribution Percentage

Costs billed by BNSF to CARB were in excess of CARB's share of the actual costs incurred by BNSF's subcontractor. The contract between CARB and BNSF states that CARB agreed to reimburse BNSF for actual expenditures incurred in accordance with the rates specified in the contract. The contract provided for a cost of $1,430,000 for each locomotive, based on a price quote received from the National Railway Equipment Company (NREC). The contract also included a BNSF contribution percentage that ranged from 35 to 50 percent. Specifically, BNSF agreed to pay 50 percent of the costs for six locomotives, 40 percent of the costs for one locomotive, and 35 percent of the costs for four locomotives. Based on these contract terms, total BNSF contribution is $6,864,000 and CARB or federal contribution is $8,866,000.

Once the work started, BNSF billed CARB for costs based on invoices submitted by NREC for each locomotive. BNSF also billed CARB for materials that BNSF provided to NREC (at no cost to the subcontractor) for use in the construction of the locomotives. BNSF billed NREC's costs using the contract-required contribution percentages. However, BNSF did not invoice CARB for the BNSF-provided materials using the contract-required contribution percentages. In addition, BNSF invoiced CARB for a remote-control system that was not for the locomotives repowered under the cooperative agreement.

We calculated the total cost for each locomotive based on the invoices submitted by NREC and BNSF. We limited eligible costs for each locomotive to the $1,430,000 rate noted in the contract between CARB and BNSF. The federal share we calculated is shown in table 1.

Table 1: OIG's calculation of federal share

Locomotive	Total cost	Eligible cost per contract	Federal percentage of eligible costs	Federal share
1293	$1,502,111	$1,430,000	50%	$715,000
1294	1,505,807	1,430,000	50%	715,000
1300	1,414,206	1,414,206	50%	707,103
1301	1,415,209	1,415,209	65%	919,886
1302	1,394,614	1,394,614	60%	836,768
1303	1,423,281	1,423,281	65%	925,133
1304	1,417,316	1,417,316	65%	921,255
1305	1,414,495	1,414,495	65%	919,422
1306	1,412,752	1,412,752	50%	706,376
1307	1,411,838	1,411,838	50%	705,919
1308	1,400,058	1,400,058	50%	700,029
	$15,711,687			**$8,771,891**

Source: Total costs are from invoices submitted by NREC and BNSF. Federal percentages are from the contract between CARB and BNSF. Eligible cost per contract and the federal share are based on the OIG's analysis of the data.

The costs claimed and questioned are summarized below:

Table 2: Summary of questioned costs

Cost category	Amount
Total project costs [a]	$15,711,687
Amount claimed [b]	8,866,000
Federal share based on contribution share per locomotive [c]	8,771,891
Cumulative cash draw [d]	8,866,000
Amount due EPA [c]	**$94,109**

Source: See notes below.
   a. Total project costs amount is based on invoices BNSF submitted to CARB.
   b. Amount claimed is from the final federal financial report that CARB submitted to the EPA under the cooperative agreement.
   c. Federal share and amount due EPA are based on the OIG's analysis of the data provided by CARB.
   d. Cumulative cash draw is from EPA and CARB's accounting records.

CARB staff agreed with the total costs and percentages provided in table 1 above, but believed the intent of the contract with BNSF was to reimburse BNSF for total project costs without a ceiling amount for each locomotive. CARB staff said the rate cited in the contract refers to the contribution percentage, not the cost per locomotive. CARB believes the calculation for eligible costs should be based on the total cost of the locomotive without the $1,430,000 ceiling.

Conclusion

The OIG's analysis determined that eligible costs for each locomotive were limited to the contract cost of $1,430,000 agreed to by CARB and BNSF. CARB staff said the intent of the contract was to apply CARB's contribution percentage to the total cost of the individual locomotives, regardless of whether the cost exceeded $1,430,000. However, the contract does not specifically indicate that the rates only apply to contribution percentages. Therefore, we believe the rates in the contract refer to both the cost ceiling and the contribution percentage for each locomotive. We believe $94,109 is ineligible.

Recommendations

We recommend that the Region 9 Regional Administrator:

1. Recover the ineligible amount of $94,109 claimed under the cooperative agreement from CARB.

2. Require CARB to establish internal controls prior to any future awards, to ensure that contractor billings comply with contract terms and conditions.

EPA and Recipient Comments

The OIG received comments on the draft report from Region 9 and CARB, as shown in appendices A and B, respectively. Region 9 and CARB also provided supplemental documentation as support for their comments, and although not included in this report the supplemental documentation is available upon request.

Region 9 agreed with the recommendations and said that it will work with CARB to ensure that contractor billings comply with EPA requirements and contract terms and conditions. CARB acknowledged that BNSF erroneously billed a remote control system that was not used for the locomotives repowered under the project. However, CARB disagreed that $94,109 of the project costs is ineligible. CARB said the issue is whether the contract ceiling is on each locomotive or on the total projects. CARB believes it should be given the discretion to determine the intent and interpretation of the contract provisions and managed the contract based on a not-to-exceed total project cost ceiling of $8,866,000. However, CARB acknowledges the basis for the questioned costs and the OIG's rationale.

OIG Response

The OIG's position remains unchanged. Region 9 concurred with our recommendations. Although CARB disagreed that the $94,109 is ineligible, CARB repaid the $94,109 to the EPA on January 9, 2014. We acknowledge the repayment and consider recommendation 1 resolved.

Chapter 4
Noncompliance With Laws, Regulations and Cooperative Agreement Conditions

As part of our examination, we reviewed CARB's compliance with applicable laws, regulations, and terms and conditions of the cooperative agreement. We found that CARB and BNSF have not scrapped or remanufactured old engines taken from the repowered locomotives in accordance with the cooperative agreement and Energy Policy Act requirements because CARB allowed BNSF the option to ban the old engines from operating in California as an alternative to scrapping. CARB said it did not require BNSF to scrap or remanufacture the old engines because of the unique nature of locomotives. CARB also believed the ban option was approved by the EPA. The potential use of the old engines by BNSF outside of California could offset emissions reductions gained by the newer engines and result in no net environmental benefit being derived from the project. As a result, we questioned the $8,866,000 claimed under the cooperative agreement. Since $94,109 of the $8,866,000 claimed has already been questioned in chapter 3 of this report, we are only questioning here the remaining $8,771,891.

In addition, CARB did not report jobs created or retained in accordance with Office of Management and Budget (OMB) guidance. CARB reported jobs created based on total hours expended on the project. OMB guidance requires that jobs funded partially with Recovery Act funds only be counted based on the proportion funded by the Recovery Act. During the course of the audit, CARB recalculated the jobs created and agreed to work with EPA Region 9 to revise the number of jobs reported.

CARB Did Not Comply With Requirement to Scrap or Remanufacture

CARB did not comply with requirements of the cooperative agreement and Energy Policy Act to scrap or remanufacture the older locomotive engines. CARB allowed BNSF the option to ban the use of the old engines in California as an alternative because of the unique nature of locomotives. However, this option is not provided for by the Energy Policy Act or the cooperative agreement.

CARB Provided Alternative to Scrapping and/or Remanufacturing

CARB's contract with BNSF to repower switch-yard locomotives provided the option to ban or scrap older, existing engines from California operations. The contract states that to use the ban option, the engines must be removed from older, existing locomotives. The railroad company must also make sure that the original older engines will not operate in California. The contract does not specifically require remanufacturing of older engines under the ban option.

In the project final report, CARB reported that BNSF retained possession of the engines taken from the repowered locomotives. The report states that the engines will be upgraded to the best possible emissions and performance standards as the need arises. The engines will be placed in locomotives that will be banned from operating in California. CARB stated that BNSF will notify CARB if an older engine is placed in another locomotive. CARB can then track the locomotive to ensure the locomotive is not observed operating in California during rail yard inspections, field surveys or photographic tracking.

BNSF Requested Ban Option

CARB staff said BNSF raised concerns about the lack of flexibility when it comes to provisions for the scrappage and remanufacture of older engines. BNSF requested that CARB provide the option to ban the engines from California, which is consistent with CARB's other funding programs using incentives (e.g., Proposition 1B and the Carl Moyer Program).

CARB believes that unique engine costs, the need to reuse engines for rebuilds, and the ability to track locomotive operations provides the basis for the option to ban older engines from California instead of scrapping or remanufacturing them. CARB staff stated that due to the cost and size of locomotive engines, the engines that are sent to scrap yards are typically overhauled or remanufactured. CARB believes that banning the engines was an acceptable scrapping method in accordance with the language of programmatic condition 10 of the cooperative agreement.

CARB Believed EPA Region 9 Approved the Ban Option

CARB believes that EPA Region 9 provided oral approval for the option to ban the locomotives and that the approval is consistent with the programmatic language of the cooperative agreement. CARB staff also said that the EPA had opportunities to review CARB's proposed ban option in its request for proposal and contract with BNSF but the EPA did not question the option.

EPA Region 9 staff did not recall approving the ban option. Region 9 staff said the options are well defined in the cooperative agreement and banning locomotives is not an option.

Energy Policy Act and Cooperative Agreement Do Not Provide Ban Option

Neither the Energy Policy Act nor the cooperative agreement allows for the option to ban the use of older engines as described by CARB. The Energy Policy Act, Title VII, subtitle G, section 793(d)(3), allows use of grant funds for certified engine configurations. Section 791(2) requires that in the case of an engine configuration replacing an existing engine, the replaced engine should be returned to the supplier for remanufacturing to a more stringent set of engine emissions or be designated as scrappage. Programmatic condition 10 of the cooperative agreement requires that the engine being replaced be scrapped within 90 days of the replacement or be returned to the original engine manufacture for remanufacturing to a cleaner standard.

Programmatic condition 10 provides for other acceptable scrapping methods with EPA approval. However, the condition defines scrappage "as a permanently disabled engine or vehicle, no longer suitable for use." CARB's option to ban the older engines did not result in permanently disabling the engines. As explained in the second paragraph of the subsection "CARB Provided Alternative to Scrapping and/or Remanufacturing," BNSF has not permanently disabled the engines but retains possession with the possibility of using the engines at a later date.

Conclusion

CARB and BNSF have not scrapped or remanufactured the engines taken from the repowered locomotives in accordance with the cooperative agreement or Energy Policy Act requirements. The ban option CARB allowed is not compliant with the Energy Policy Act or the cooperative agreement. CARB staff stated that the cooperative agreement did not provide a specific time frame for the remanufacture. In our opinion, although the cooperative agreement did not provide a specific time frame for remanufacturing the old engines, the cooperative agreement stated that to be considered a repower, the purchase of new engines must be accompanied by the scrappage or remanufacture of the old engines. In addition, the cooperative agreement states that evidence of appropriate disposal is required in a final assistance agreement report submitted to the EPA. We believe that the cooperative agreement requirements express intent to complete remanufacture prior to close-out of the cooperative agreement.

In response to the audit, CARB and BNSF signed a written agreement on November 18, 2013, to scrap or remanufacture the 11 older locomotive engines within 18 months of the agreement date. If there is a delay in the scrap or remanufacturing process due to unforeseen circumstances, BNSF agreed to inform CARB in writing the reason for the delay at least 90 days prior to the end date of the scrap/remanufacture. BNSF may request a time extension in writing with the consent of all signatories.

CARB Jobs Report Did Not Comply With OMB Requirements

CARB incorrectly reported the number of jobs created and retained under the Recovery Act. Recovery Act Section 1512 requires each recipient receiving Recovery Act funds from a federal agency to submit a quarterly report with an estimate of the number of jobs created and the number of jobs retained by the project. OMB issued various guidance documents to implement Recovery Act requirements. On December 18, 2009, OMB issued guidance M-10-8 to update, among other things, the method for estimating the number of jobs created and retained. The guidance defines jobs created or retained as those jobs funded in the quarter by the Recovery Act. Jobs funded partially with Recovery Act funds will only be counted based on the proportion funded by the Recovery Act.

CARB reported the number of jobs created using total hours NREC worked on the locomotives. Recovery Act funds only provided approximately 56 percent of the funding for the project. As a result, CARB overstated the number of jobs reported. CARB recalculated the jobs created and has agreed to work with the EPA to revise the number of jobs reported.

Recommendations

We recommend that the Region 9 Regional Administrator:

3. Verify that CARB and BNSF scrap or remanufacture the old engines in accordance with the November 18, 2013, agreement.

4. Require CARB to provide documentation that the replaced engines were scrapped or remanufactured in accordance with the November 18, 2013, agreement.

5. Recover the federal share of $8,771,891 claimed in the event CARB does not scrap or remanufacture the engines in accordance with November 18, 2013, agreement.

6. Require CARB to revise the jobs reported as created or retained to reflect the number of jobs funded by the Recovery Act.

EPA and Recipient Comments

Region 9 concurred with the recommendations. CARB concurred with recommendation 6, but did not state whether it concurred with recommendations 3 to 5. CARB reiterated that it was clear in its intent to allow BNSF to ban the engines from operation in California and that the cooperative agreement and the Energy Policy Act provided the flexibility to include alternatives to scrappage. In the draft report, we recommended that the Region 9 Regional Administrator require CARB to scrap or remanufacture the replaced engines in accordance with the Energy Policy Act and the terms and conditions of the cooperative agreement. In response to the recommendation, CARB and BNSF signed a written agreement on November 18, 2013, to scrap or remanufacture the 11 older locomotive engines within 18 months of the agreement date. If there is a delay in the scrap or remanufacturing process due to unforeseen circumstances, BNSF agreed to inform CARB in writing the reason for the delay at least 90 days prior to the end date of the scrap/remanufacture. BNSF may request a time extension in writing with the consent of all signatories. Region 9 and EPA headquarters support the agreement. Region 9 said it will continue to work with CARB to track progress. Region 9 said the EPA does not anticipate the need to recover $8,771,891 based on the terms of the agreement between CARB and BNSF. However, if BNSF does not properly scrap or remanufacture the locomotives in compliance with the terms and conditions of the agreement, Region 9 will seek to recover grant funds from CARB.

To address recommendation 6, CARB provided the revised number of jobs created and retained and Region 9 said it is in the process of reporting revised numbers.

OIG Response

We acknowledge the November 18, 2013, written agreement from BNSF to scrap or remanufacture the locomotive. We agree with Region 9's plan to track progress until the recommendations are fully resolved. In the event CARB or BNSF does not scrap or remanufacture the engines as required by the agreement, Region 9 should recover the $8,771,891 questioned.

Based on our research, Recovery Act Section 1512 quarterly reports for prior periods cannot be amended. Since CARB has submitted the revised report and explanations for the error, we consider recommendation 6 resolved.

Chapter 5
CARB Did Not Provide Actual
Emissions Reduction Calculations

CARB did not provide actual emission benefit calculations as required by the cooperative agreement. CARB staff said emissions reductions are based on estimates of fuel usage because Class I railroads do not calculate or estimate individual locomotive fuel consumption. As a result, CARB does not have reasonable assurance that the repowered locomotives are achieving projected emissions reductions and human-health benefits. Further, DERA program results may be overstated or understated.

CARB Calculated Emissions Reductions Using Estimated Fuel Usage

For the final report, CARB calculated a range of emissions reductions for the repowered locomotives based on estimated annual fuel usage and EPA-certified emission factors for NOx and particulate matter. CARB used the emission calculation methodology from California's Carl Moyer Program criteria and guidelines. According to Region 9, the Carl Moyer guidelines are the EPA's guidelines. The Carl Moyer methodology allows emissions reductions to be calculated based on hours of operation, fuel consumption or miles traveled. CARB used the fuel-consumption method, which uses the gallons per year and emission factors, to determine the emissions reductions.

CARB staff stated they used several methods to determine an annual fuel-consumption rate for switch-yard locomotive engines. For older engines, CARB used the hourly fuel rate and estimated annual operation hours. CARB also estimated annual fuel usage based on a 2004 inventory of intrastate locomotives and the average fuel consumed. CARB estimated fuel usage for newer switch-yard locomotives using a demonstration study of a Tier 4 locomotive in the South Coast Air Basin. CARB also used fuel-consumption data provided by BNSF for switch-yard locomotives operating in Texas. Based on these data, CARB's estimated fuel usage ranges from 40,000 to 50,000 gallons per year for older locomotives and 30,000 to 40,000 gallons per year for newer locomotives.

For the project final report, CARB used a range of potential reductions for NOx and particulate matter based on the fuel usage estimates noted above. Estimated NOx emissions ranged from 0.047 to 0.026 tons per day. CARB's estimated particulate matter emissions ranged from 0.002 to 0.0011 tons per day. The difference between the high end of the range and the low end represents a 45-percent variance. Table 3 summarizes the emissions reductions calculations.

Table 3: CARB's calculation of potential emissions reductions

	High end of range	Low end of range	Variance
Gallons per year (old engines)	50,000	40,000	
Gallons per year (new engines)	40,000	30,000	
NOx emissions (tons/day/locomotive)	0.047	0.026	45%
Particulate emissions (tons/day)	0.002	0.0011	45%

Source: Project final report submitted by CARB.

The cooperative agreement was expected to provide potential reductions of approximately 0.045 and 0.0018 tons per day of NOx and particulate matter, respectively. However, as illustrated by the data provided by CARB, fuel usage varies significantly and the project may not achieve the emissions reduction target.

The cooperative agreement states the recipient will include "actual" emissions benefit calculations in the final report. However, CARB has not required BNSF to provide actual fuel consumption for the 11 locomotives repowered under the cooperative agreement. CARB stated that Class I railroads do not calculate or estimate individual locomotive diesel-fuel consumption in normal operations because diesel-fuel consumption is normally calculated on a national basis. BNSF's proposals stated that fuel records for currently operating switch locomotives in its rail yards are not maintained. CARB staff said it is not realistic to require BNSF to provide actual fuel usage, and the staff also believes that the use of estimates provides an adequate surrogate for actual data.

Conclusion

CARB calculated potential emissions reductions for the repowered locomotives based on estimates of fuel usage because actual fuel-usage data for individual locomotives is not available. CARB provided a range of emissions reductions based on estimated fuel usage. Unless CARB can provide actual fuel usage, the EPA does not have reasonable assurance that the project will achieve projected emissions reductions or expected environmental results and human-health benefits.

Recommendations

We recommend that the Region 9 Regional Administrator:

7. Work with CARB to start collecting actual fuel usage data and develop a more accurate calculation of project results based on actual fuel usage.

8. Adjust DERA program reporting of CARB project results to reflect recalculated results.

EPA and Recipient Comments

Region 9 and CARB disagreed with the finding and recommendations. Region 9 said CARB provided the emission reduction calculations based on the most accurate data available. CARB said it followed the industry standard practice by using the EPA's emission reduction calculation methodology, which relies on fuel consumption estimates and EPA-certified emission factors for NOx and particulate matter.

Region 9 explained that the EPA's emission quantification models used various assumptions to generate the emission factors for locomotive engines. Region 9 further said that estimates, not actual or quantifiable emission, are used in the EPA's engine rules as well as the tool used for quantifying the emissions reductions for DERA projects. Region 9 and CARB said that railroads rarely track or retain data on fuel use for switcher locomotives because it represents a small percentage of total Class I locomotive fuel use. CARB said this tracking is only done as a part of a funded technology demonstration project and was unavailable for the cooperative agreement. Region 9 said the U.S. Department of Transportation does not require reporting annual fuel use for each locomotive. Region 9 and CARB said the OIG's recommendation is not feasible because requiring railroads to track and report fuel use data would be very time consuming and costly.

CARB said it relied on fuel use estimates from multiple correlative data sources that represented the best available information. CARB said it sought out and updated that data throughout the project to ensure the emission reduction estimates from the project were the most accurate possible. As a result, CARB said it does have assurance that the repowered locomotives are, in fact, achieving the range of projected emission reductions and health benefits.

OIG Response

The OIG's position remains unchanged. The OIG has already addressed CARB's calculation methodology in the draft report. The OIG does not question the emission reduction calculation methodology or the use of estimated emission factors for NOx or particulate matter. Our issue is the use of estimated, rather than actual, fuel usage. We acknowledge that actual usage data does not currently exist and collecting actual usage data may be costly. However, "actual" emissions benefit calculations is expressly required under the cooperative agreement and BNSF can collect this data. It collected this data for the locomotives operating in Texas. While CARB stated that it has assurance that the repowered locomotives are achieving the range of projected reduction and health benefits. However, the range varies significantly. The variance between the high end and low end of the range is approximately 45 percent. Unless CARB can provide actual fuel usage, the EPA and the public do not have reasonable assurance that the project will achieve projected emissions reductions or expected environmental results and human-health benefits.

Status of Recommendations and Potential Monetary Benefits

		RECOMMENDATIONS				POTENTIAL MONETARY BENEFITS (in $000s)	
Rec. No.	Page No.	Subject	Status[1]	Action Official	Planned Completion Date	Claimed Amount	Agreed-To Amount
1	8	Recover the ineligible amount of $94,109 claimed under the cooperative agreement from CARB.	C	Region 9 Regional Administrator	12/31/13	$94	$94
2	8	Require CARB to establish internal controls prior to any future awards, to ensure that contractor billings comply with contract terms and conditions.	U	Region 9 Regional Administrator			
3	12	Verify that CARB and BNSF scrap or remanufacture the old engines in accordance with the November 18, 2013, agreement.	O	Region 9 Regional Administrator	5/18/15		
4	12	Require CARB to provide documentation that the replaced engines were scrapped or remanufactured in accordance with the November 18, 2013, agreement.	O	Region 9 Regional Administrator	5/18/15		
5	12	Recover the federal share of $8,771,891 claimed in the event CARB does not scrap or remanufacture the engines in accordance with the November 18, 2013, agreement.	O	Region 9 Regional Administrator	5/18/15	$8,772	
6	12	Require CARB to revise the jobs reported as created or retained to reflect the number of jobs funded by the Recovery Act.	C	Region 9 Regional Administrator	12/31/13		
7	15	Work with CARB to start collecting actual fuel usage data and develop a more accurate calculation of project results based on actual fuel usage.	U	Region 9 Regional Administrator			
8	15	Adjust DERA program reporting of CARB project results to reflect recalculated results.	U	Region 9 Regional Administrator			

[1] O = Recommendation is open with agreed-to corrective actions pending.
 C = Recommendation is closed with all agreed-to actions completed.
 U = Recommendation is unresolved with resolution efforts in progress.

Agency's Comments on Draft Report

UNITED STATES ENVIRONMENTAL PROTECTION AGENCY
REGION 9
75 Hawthorne Street
San Francisco, CA 94105

December 9, 2013

MEMORANDUM

SUBJECT: Response to Office of Inspector General draft report Project No. OA-FY13-0210
"Weaknesses Disclosed in California Air Resource Board's Compliance with
Laws, Regulations and Recovery Act Requirements," September 25, 2013

FROM: Thomas McCullough
Assistant Regional Administrator

TO: Robert Adachi
Director, Forensic Audits
Office of the Inspector General

Thank you for the opportunity to respond to the issues and recommendations in the subject audit
report. Following is a summary of the Environmental Protection Agency Region 9's (Region 9)
overall position, along with our position on each of the report recommendations. For those
recommendations with which Region 9 agrees, we have provided high-level intended corrective
actions and estimated completion dates to the extent we can. For those report recommendations
with which Region 9 does not agree, we have explained our position and proposed alternatives to
the recommendations. For your consideration, we have included a Technical and Substantive
Comments Attachment 1 to supplement this response.

EPA REGION 9'S OVERALL POSITION
Region 9 shares your interest in ensuring we protect the integrity of our cooperative agreements
as we achieve environmental goals. We agree with most of the draft report recommendations,
and have noted the two recommendations with which we disagree. In addition, we have general
comments and recommended changes which are noted in the Technical and Substantive
Comments Attachment. For example, we propose the report title be revised with more neutral
wording to more accurately reflect the findings in the context of the positive results achieved.
We have moved forward to undertake many of the recommendations raised in your report, as
outlined below, and look forward to addressing and resolving other report recommendations.

AGENCY'S RESPONSE TO REPORT RECOMMENDATIONS

Agreements

No.	Recommendation	High-Level Intended Corrective Actions	Estimated Completion
1	Recover the ineligible amount of $94,109 claimed under the cooperative agreement	Region 9 agrees with this recommendation and is working with the California Air Resources Board (CARB) to return $94,109, or the full amount of the ineligible funds. However, minor inaccuracies exist in this section of the report (see the Technical and Substantive Comments Attachment for rewording suggestions).	1st Quarter FY14
2	Require CalARB to establish internal controls prior to any future awards, to ensure that contract billings comply with contract terms and conditions.	Region 9 agrees with this recommendation and will work with CARB to ensure adequate internal controls are put in place to ensure billing meets EPA's requirements and terms and conditions.	2nd Quarter FY14
3	Require CalARB to scrap or remanufacture the replaced engines in accordance with the Energy Policy Act and the terms and conditions of the cooperative agreement.	Region 9 agrees with this recommendation. On November 18, 2013, EPA Region 9 received an executed Closeout Agreement between CARB and Burlington Northern Santa Fe (BNSF) which details BNSF's plan to scrap and/or remanufacture all of the 11 old locomotive engines (see Attachment 2 Closeout Agreement between CARB and BNSF). EPA Region 9 and Headquarters staff supports this Agreement and will continue to work with CARB to track progress.	3rd Quarter FY15 or no later than May 18, 2015
4	Require CalARB to provide documentation that the replaced engines were scrapped or remanufactured.	See response to Recommendation 3 above. The Agreement between CARB and BNSF contains a provision which states that CARB will meet the requirements of Programmatic term and condition P.10 (Scrappage and/or Remanufacture requirement). Region 9 will work with CARB to ensure that proper scrapping and/or remanufacturing of the eleven locomotives occurs.	3rd Quarter FY15 or no later than May 18, 2015
5	Recover $8,771,891 of the remaining total federal share of the claimed costs, unless CalARB complies with the scrappage requirements.	Region 9 and Headquarters do not anticipate that they will need to recover $8,771,891 based on the terms of the Closeout Agreement referenced above. However, if BNSF does not properly scrap and/or remanufacture the engines pursuant to the terms of the aforementioned Agreement, Region 9 will seek to recover grant funds from CARB.	3rd Quarter FY15 or no later than May 18, 2015
6	Require CalARB to revise the jobs reported as created or retained to reflect the number of jobs funded by the Recovery Act.	Region 9 agrees with this recommendation. It should be noted that CARB has provided the revised jobs created and/or retained and Region 9 is in the process of reporting such.	1st Quarter FY14

Disagreements

No.	Recommendation	Agency Explanation/ Response	Proposed Alternative
7	Work with CalARB to develop a more accurate calculation of project results based on actual fuel usage.	Region 9 and Headquarters do not agree with Chapter 5 since CARB provided the most accurate emission reductions. Additional information, including the industry standards for reporting these types of results, is provided in the Technical and Substantive Comments Attachment.	Delete Chapter 5.
8	Adjust DERA program reporting of CalARB project results to reflect recalculated results.	As mentioned above, Region 9 and Headquarters do not agree with Chapter 5 since CARB provided the most accurate emission reductions. Additional information is provided in the Technical and Substantive Comments Attachment.	Delete Chapter 5.

CONTACT INFORMATION

If you have any questions regarding this response, please contact Ben Machol, Manager, Clean Energy and Climate Change Office, Air Division at (415) 972-3770 or Machol.Ben@EPA.gov.

Attachment 1: Technical and Substantive Comments
Attachment 2: Closeout Agreement with CARB and BNSF

cc:
Deborah Jordan, Director, Air Division, U.S. EPA Region 9

Ben Machol, Manager, Clean Energy and Climate Change Office, Air Division, U.S. EPA Region 9

Jack Kitowski
Assistant Chief, Stationary Source Division
CA Air Resources Board
1001 I Street
PO Box 2815
Sacramento, CA 95812

Trina Martynowicz, Environmental Protection Specialist, Air-9, Region 9

Penelope McDaniel, Environmental Protection Specialist, Air-9, Region 9

Marie Ortesi, Team Lead, Audit Follow-up, MTSD-4-2, Region 9

Magdalen Mak, Audit Follow-up Coordinator, MTSD-4-2, Region 9

Lela Wong, Auditor, OIG

Attachment 1: Technical and Substantive Comments
Response to Office of Inspector General (IG) Draft Report Project No. OA-FY13-0210 "Weaknesses Disclosed in California Air Resource Board's Compliance with Laws, Regulations and Recovery Act Requirements," September 25, 2013

EPA Region 9 and Headquarters are providing the following technical and substantive corrections to the IG draft report:

1. The project number is incorrectly referenced as "QA-FY13-0210" on the title page of the report and should be changed to "OA-FY13-0210."

2. The title of the report should be revised with more neutral wording to better reflect the findings in the context of the positive results achieved, for example: "Examination of the California Air Resource Board's Compliance with Laws, Regulations and Recovery Act Requirements."

3. A more common acronym for the California Air Resources Board is "CARB," not CalARB, therefore "CalARB" should be revised throughout the report to "CARB."

4. All sections of the report that discuss scrapping or remanufacturing the old locomotive engines should be revised to state "scrapping and/or remanufacturing," since Burlington Northern Santa Fe Railway Company (BNSF) may choose to undertake either or both of these activities.

5. CARB's original workplan included repowering eight locomotives, as mentioned in the draft report. EPA recommends the IG include additional language to accurately highlight these significant achievements. For example, in Chapter 2's Noteworthy Achievements, the following language should be added: "Because BNSF repowered three additional locomotives, significant air quality benefits were achieved by these emission reductions. The communities surrounding these rail yards will experience improved air quality due to the achievements by CARB, BNSF, and the engine manufacturer."

6. The final sentence in the third paragraph on page 6 of the report should list the correct funding amount that will be returned so that it reads: "CARB agreed with our issue on the remote-control system and is currently working with Region 9 to refund to the EPA the over-billed amount of $94,109."

7. On the top of page 8 of the report, the final sentence of the first paragraph should be changed to: "However, CARB has agreed to return the full ineligible amount of $94,109."

8. Under Recommendation number 1 on page 8 of the report, the IG recommends that the Region 9 Regional Administrator recover the ineligible amount of $94,109 claimed under the cooperative agreement. EPA Region 9 suggests adding the following information in response to that recommendation: "CARB agrees with this recommendation and will return the full ineligible amount of $94,109 to EPA Region 9."

9. Under Recommendation number 3 on page 12 of the report, the IG recommends that the Regional Administrator require CARB to scrap or remanufacture the replaced engines in accordance with the Energy Policy Act and the terms and conditions of the cooperative agreement. On November 18, 2013, EPA Region 9 received an executed Closeout Agreement between CARB and BNSF which details BNSF's plan to scrap and/or remanufacture all of the 11 old locomotive engines (see Attachment 2 Closeout Agreement with CARB and BNSF.) EPA Region 9 and Headquarters support this agreement, and will continue to work with CARB to track progress to ensure that the terms and conditions of the cooperative agreement are met. If BNSF decides to remanufacture any locomotive engine BNSF will meet EPA's standards for remanufactured engines. The final IG report should make note of the Closeout Agreement. In addition, Region 9 recommends that item number 3 on page 12 of the report be revised to read as follows: "The replaced engines will be scrapped and/or remanufactured pursuant to the terms delineated in the November 18, 2013 Closeout Agreement between CARB and BNSF." EPA Region 9 and Headquarters has highly encouraged the old engines to be scrapped and/or remanufactured within one year of the signed date of this Agreement, though Region 9 agreed to allow these activities to occur within 18 months, or before May 18, 2015.

10. Under Recommendation numbers 7 and 8 on page 14 of the report, the IG recommends that EPA work with CARB to develop a more accurate calculation of project results based on actual fuel usage and adjust the DERA program reporting to reflect these recalculated results. EPA Region 9 and Headquarters disagrees with the IG's recommendations and fully supports CARB's emission reduction calculation methodology, which is based on the most accurate and available information. CARB regularly updates their methodology and calculations for emission reductions and has extensive experience calculating emission reductions from in-use and currently operating locomotives. CARB provides approximately $90 million annually in grants for diesel emission reductions activities through their Carl Moyer Program, including switcher locomotives, similar to EPA's DERA program and this specific cooperative agreement.[1] Therefore, CARB has a strong need to and has extensive past experience in accurately calculate emission reductions from diesel projects. Additional information on the way in which EPA and CARB calculate overall emission reductions, as well as typical industry practice for quantifying fuel use are provided below.

EPA's emission quantification models, including those used for setting emission standards for manufacturing locomotive engines, use various assumptions to generate the emission factors. Estimates, not actual or quantifiable emissions are used in EPA's engine rules, as well as EPA's Diesel Emission Quantifier (DEQ), the tool used for quantifying the emission reductions for Diesel Emission Reduction Act (DERA) projects. As the DEQ website states, this tool provides estimates of, not actual emission reductions.[2]

---

[1] CARB's "Carl Moyer Memorial Air Quality Standards Attainment Program," http://www.arb.ca.gov/msprog/moyer/moyer.htm.
[2] EPA Nation Clean Diesel Campaign's "Diesel Emission Quantifier," http://www.epa.gov/cleandiesel/quantifier/.

Because switcher locomotives represent such a small percent of the Class 1 locomotive fleet's fuel use and the cost to maintain such records is so high, the railroads rarely track or retain this data. Therefore, fuel usage estimates are used to quantify emission reductions. EPA's methodology for calculating emission reductions describes how the agency generates estimated, but never actual emission rates.[3]

The common Class 1 railroad practice of tracking and reporting fuel usage differs greatly from other DERA recipients, such as long-haul truck or school bus fleets. Railroad companies do not track nor regularly quantify the actual fuel consumed for each locomotive. Unlike diesel-fueled trucks or school buses that fuel at a specific fueling station owned by a third party, railroads have their own fueling stations located at the rail yard for both switcher and line-haul locomotives.

Department of Transportation's (DOT) Surface Transportation Board requires all Class 1 railroads to report quarterly and annual fuel use for both switcher locomotives, which this cooperative agreement funded, as well as line-haul locomotives.[4] DOT does not mandate reporting annual fuel use for each locomotive. BNSF switcher locomotives only consumed approximately 3.5% of the total fuel used for all BNSF locomotives in 2012.[5] Railroads purchase their fuel in bulk and quantify the amount of fuel used for all locomotives, hardly ever for an individual locomotive. In addition, it is not common industry practice for Class 1 railroads to monitor fuel consumption on a data log for a given switcher locomotive due to the high cost of tracking, monitoring and reporting.

Meeting the IG's recommendation is not feasible. Requiring railroads to track and report fuel use data would be very time consuming and costly and seen as a heavy administrative burden for the old dirtier and/or new cleaner locomotives. Based on common industry practices and the way in which EPA quantifies emission factors for engines used in locomotives, CARB provided emission reduction calculations that are acceptable. EPA Region 9 and Headquarters believes the emission reductions CARB provided are the most accurate and available, therefore Chapter 5 of this draft report should be deleted altogether. If the IG does not agree with Region 9's conclusion to delete this section of the report, the Region will work with the IG to provide changes to ensure the accuracy of this chapter.

---

[3] EPA Office of Transportation and Air Quality's "Emission Factors for Locomotives," April 2009, http://www.epa.gov/otaq/regs/nonroad/locomotv/420f09025.pdf.

[4] DOT Surface Transportation Board's "Annual Reports R-1 Selected Schedules and Complete Annual Reports," http://www.stb.dot.gov/stb/industry/econ_reports.html.

[5] BNSF's "Class I Railroad Annual Report Restatement To The Surface Transportation Board For the Year Ending December 31, 2012," Page 91 "750. Consumption of Diesel Fuel," http://www.bnsf.com/about-bnsf/financial-information/surface-transportation-board-reports/pdf/12R1.pdf.

CARB's Comments on Draft Report

Air Resources Board

Mary D. Nichols, Chairman
1001 I Street • P.O. Box 2815
Sacramento, California 95812 • www.arb.ca.gov

Matthew Rodriquez
*Secretary for
Environmental Protection*

Edmund G. Brown Jr.
Governor

November 25, 2013

Mr. Robert Adachi
Director, Forensic Audits
U.S. Environmental Protection Agency
Office of the Inspector General
75 Hawthorne Street
y!h Floor, M/C IGA-1
San Francisco, California 94105

Dear Mr. Adachi:

The Air Resources Board (ARB) is documenting the actions we have taken to resolve the issues raised in the draft findings and recommendations of the U.S. Environmental Protection Agency (U.S. EPA) Office of Inspector General (OIG) audit report number OA-FY13-0210. The OIG audit covers the American Recovery and Reinvestment Act (ARRA) project award of $8.88 million to ARB under cooperative agreement 2A-OOT13801. The project achieved its objective to cut air pollution and health risk near railyards in Southern California through incentives for BNSF Railway to replace old locomotive engines with cleaner models.

Although we do not concur with the majority of the OIG conclusions, we appreciate that the discussion in the draft report accurately characterizes the information we provided to the auditors. ARB has addressed all of the issues raised by the OIG during the course of this audit. This letter provides documentation and our commitment to follow through with U.S. EPA to ensure the OIG concerns are fully resolved. I would like to specifically highlight an edit we are requesting to the title of the draft audit report to be more consistent with other published OIG reports- replacing "Weakness Disclosed in..." with "Examination of..." in the title. Thank you for considering this edit. In addition to this letter, please find copies of the following supporting documents enclosed:

- Background and ARB's responses to each of the draft findings of the OIG audit.
- The signed legal agreement between BNSF and ARB that requires BNSF to scrap or remanufacture eleven older BNSF locomotive engines by May 18, 2013.
- Documentation of payment to ARB of $94,109 by BNSF on November 18, 2013.
- ARB letter to U.S. EPA Region 9 addressing several audit issues.
- The U.S. EPA OIG draft audit report with ARB's suggested updates and factual corrections identified in redline/strikeout format.
- ARB's signed management representation letter.

Mr. Robert Adachi, Director
November 25, 2013
Page 2

We remain convinced that the ARRA co-funded replacement of old engines with new genset models in eleven locomotives was a successful, cost-effective project with on-time delivery that continues to produce real emission reductions for communities near Southern California railyards.

ARB appreciates the opportunity to provide the U.S. EPA OIG auditor with written responses to the draft audit. If you have any questions, please contact Mr. Jack Kitowski, Assistant Chief, Stationary Source Division at (916) 445-6102 or jkitowsk@arb.ca.gov.

Sincerely,

ti),

Richard W. Corey
Executive Officer

Enclosures

cc: Ms. Lela Wong, CPA, CFE
       Project Manager, Office of Inspector General
       U.S. Environmental Protection Agency
       75 Hawthorne Street
       San Francisco, California 94105

       Ms. Deborah Jordan, Director
       Air Division, Region 9
       U.S. Environmental Protection Agency
       75 Hawthorne Street
       San Francisco, California 94105

       Mr. Jack Kitowski
       Assistant Chief
       Stationary Source Division

RESPONSES TO DRAFT FINDINGS

I.   BACKGROUND

On July 10, 2009, U.S. EPA awarded ARB $8.88 million in ARRA funding under cooperative agreement 2A-00T13801. The OIG audit project number is OA-FY13-0210. The purpose of the ARRA award was to cut railyard locomotive emissions in the Basin through the repower or replacement of old engines in eight (8) switch locomotives with much cleaner engines using "genset" technology.

Genset switch locomotives, with the required use of ARB diesel, can reduce older switch locomotive diesel particulate matter (PM) and oxides of nitrogen (NOx) emissions by up to 90 percent, and significantly reduce the associated cancer risks, in and around California railyards.

On October 1, 2009, ARB released a Request for Proposal (RFP) for qualifying railroads in the Basin to compete for the ARRA funding. ARB received two proposals in response to the ARB ARRA RFP from California's two Class I railroads – BNSF and Union Pacific Railroad (UP).

ARB awarded the project to BNSF at a total project cost of $15.73 million. BNSF matched $8.87 million of ARRA funds, which was less than the U.S. EPA ARRA award of $8.88M, with $6.86 million in private funding and proposed to repower a total of eleven locomotives because of a lower per locomotive cost estimate from the genset locomotive manufacturer National Railway Equipment Company (NREC).

This project successfully repowered eleven locomotives by the ARRA project deadline of September 30, 2010, delivering significant health benefits from day one. This was accomplished by BNSF and NREC prioritizing and completing the genset switch locomotive production, testing, and delivery within three months – a process that normally takes 12 to 18 months. We understand that this accomplishment makes the project part of the five percent of all ARRA projects nationwide that were completed on time.

II.   ARB RESPONSES TO DRAFT OIG REPORT ITEMS

A   Chapter 3: Costs Claimed Were Not in Accordance With Contract Provisions

    1. Summary. The OIG identified two specific issues relating to the contractor's billings. The first issue is the contractor's erroneous billing of a locomotive control system. The second issue is whether or not the project billing is subject to a per locomotive cost cap, or a total project cost cap, for federal funding. When taken together, these issues result in the OIG recommendation that U.S.EPA Region 9 recover an ineligible amount of $94,109. While ARB disagrees with the OIG's conclusion that $94,109 of project cost is ineligible, we acknowledge the basis for the OIG's rationale. We have invoiced and received payment from BNSF for the full amount of $94,109 (see enclosed).

2. *Discussion.* ARB concurs that BNSF erroneously invoiced ARB for a remote control system that was not used for the genset switch locomotives repowered under the ARRA-funded project. This issue was acknowledged early in our discussions with U.S. EPA Region 9 and the OIG. The second issue focused on whether or not the provisions within the ARB/BNSF contract applied a "per locomotive" cost cap or a "total project" cost cap for the federal share. While ARB is clearly in compliance with the U.S. EPA terms and conditions, the difference comes in the interpretation of the ARB/BNSF contract. ARB believes it should be given discretion as to the intent and interpretation of the ARB BNSF contract provisions.

As discussed in the draft audit report, ARB staff managed the ARB/BNSF contract based on a not-to-exceed "total project" cost cap of $8,866,000, and specific per locomotive cost share percentages.

ARB staff and BNSF regarded the reference in the contract to $1.43 million per genset switch locomotive to be an estimate, with the understanding that each genset locomotive would have unique design or mechanical differences (e.g., different number of traction motors, need for Remote Control Locomotive (RCL) devices, etc.) that would result in differences in actual costs. Thus, the total cost for each of the eleven genset switch locomotives would "average" about $1.43 million, but the ARRA contribution in total could not exceed $8,866,000.

The $1.43 million per locomotive cost was clearly identified as an estimate in the April 2009 ARB ARRA application and in the October 2009 BNSF proposal provided in response to the ARB RFP. However, the tables in the ARB/BNSF contract did not include the word "estimate," which led the OIG to conclude that there was a $1.43 million per locomotive cap.

Based on the OIG auditor's interpretation that a per locomotive cost cap applies, ARB staff agrees that there is a difference of $94,109, as compared to allowing a federal share of total projects costs of up to $8.866 million. To implement the OIG auditor's interpretation, ARB staff sent an invoice to BNSF for the full $94,109 difference on October 18, 2013.

We would like to note, however, that ARB previously denied BNSF's requests for consideration of other eligible costs for nine of the genset switch locomotives with costs below $1.43 million each. The other eligible costs that were incurred by BNSF during the project were for locomotive paint and transportation. ARB denied BNSF's request for consideration of these other eligible costs because, at that time, those costs would have exceeded the total project cost share cap of $8.87 million. Had the project expenditures been tracked and approved based on a per locomotive cap, these other eligible expenses would have more than offset the $94,109. U.S. EPA Region 9 has informed us that since the project file is closed, there is no longer opportunity to provide additional documentation of these other eligible costs.

B. Chapter 4: Noncompliance With Laws, Regulations and Cooperative Agreement Conditions

1. *Summary.* The first issue is related to the requirement in the agreement between U.S. EPA and ARB that the old locomotive engines be scrapped, remanufactured, or

an alternative approved by U.S. EPA. The second issue is related to the scope of the job creation estimates required by ARRA.

As to the first issue, ARB was clear and transparent in its intent to allow BNSF to ban the old engines from operation in California as an alternative to scrappage, consistent with ARB's approach on State incentive programs. This was premised on our mistaken belief that U.S. EPA staff supported the approach. We acknowledge that we should have put our approach in writing and requested written U.S. EPA approval to be certain all parties had the same understanding. ARB and BNSF have since signed a legal agreement that requires BNSF to scrap or remanufacture the old engines within 18 months to address the draft finding in the audit.

On the second issue, ARB concurs that the job creation estimates provided by BNSF and reported by ARB inadvertently included all of the job creation benefits associated with the project, rather than the prorated benefits attributable to only the federally- funded portion of the project cost. ARB has reported the corrected numbers to U.S. EPA.

       2. Discussion. The draft OIG report questions ARB's compliance with the terms and conditions of the ARRA cooperative agreement related to the scrap or remanufacture of the older locomotive engines. While the language of the cooperative agreement and the Energy Policy Act provides for the flexibility to include alternatives to scrappage, the OIG disagrees that the option to ban the older locomotive engines from operations in California is allowed under this flexibility.

As documented by the OIG report, ARB included the ban option based on a verbal discussion with U.S. EPA Region 9 staff, after which ARB staff had the impression the ban option was acceptable. ARB was transparent about this approach – the ban option was clearly stated in the RFP for the project, and in the contract with BNSF. Both of these documents were provided to Region 9 staff for advance review, but no issues were raised about allowing this option.

However, the OIG has clarified that even if U.S. EPA had provided formal approval, the ban option was not allowed under the Energy Policy Act, and would still have been subject to an audit finding.

In response to this issue, ARB and BNSF have signed a written agreement committing BNSF to scrap or remanufacture the eleven older locomotive engines within 18 months. A copy of the agreement is attached to this letter. ARB has also forwarded a copy of the signed agreement to U.S. EPA Region 9. When BNSF completes the scrap and/or remanufacture, ARB will forward the appropriate documentation to U.S. EPA to demonstrate compliance.

For the second issue in Chapter 4, the updated estimates of the jobs created by the ARRA funded portion of the project were relayed to U.S. EPA Region 9 and are included in this package. The enclosed letter to U.S. EPA Region 9 documents ARB's formal transmittal of the updates. At this time, we understand that U.S. EPA Region 9 is working with headquarters to determine the appropriate forum to publish the revised job creation numbers since ARB can no longer access or update the project information on- line in the federal database.

C. Chapter 5: CalARB Did Not Provide Actual Emissions Reduction Calculations

1. Summary. The OIG draft report states that ARB does not have reasonable assurance that the repowered locomotives are achieving the projected emission reductions and human health benefits because actual fuel usage was not used to estimate emissions. ARB staff strongly disagrees. Actual fuel use was unavailable for this project, and is cost prohibitive for the railroads to collect on an ongoing basis for each locomotive. ARB staff relied on fuel use estimates from multiple correlative data sources that represented the best available information. ARB staff sought out and updated that data throughout the project to ensure the emission reduction estimates from the project were the most accurate possible. As a result, ARB does have assurance that the repowered locomotives are, in fact, achieving the range of projected emission reductions and health benefits.

2. Discussion. To estimate the project's emission reductions, ARB followed the industry standard practice by using U.S. EPA's emission reduction calculation methodology (http://www.epa.gov/otaq/regs/nonroad/locomotv/420f09025.pdf), which relies on fuel consumption estimates and U.S. EPA-certified emission factors for NOx and PM. Class I railroads do not typically track or measure individual locomotive diesel fuel consumption in normal operation due to the high associated cost of about $50,000 per year per locomotive. This tracking is only done as a part of a funded technology demonstration project, and was unavailable for this ARRA grant. As a result, ARB relied on several robust data sources to estimate and corroborate the switch locomotive diesel fuel consumption figures used for this project in the final report.

These data sources included information that increased the accuracy of the emission reduction estimates, but were not available to ARB in early 2009, the time of ARB's original application to U.S. EPA. These data sources included additional fuel consumption data in BNSF's grant application to ARB in late 2009, and data obtained in 2010 at the completion of a switch locomotive demonstration project.

The OIG has recommended that we work with U.S. EPA to develop a more accurate calculation of project results based on actual fuel usage. ARB recognizes the value and is committed to obtaining the best available information to estimate emission reductions.

As such, ARB has continuously strived to develop new and innovative methodologies and has continued to fund numerous technology demonstrations that provide measured fuel and emissions data. We commit to continue to work with U.S. EPA to improve on the existing calculation methodologies in order to more accurately calculate locomotive emissions.

Distribution

Regional Administrator, Region 9
Deputy Regional Administrator, Region 9
Assistant Regional Administrator, Region 9
Assistant Administrator for Air and Radiation
Director, Grants and Interagency Agreements Management Division, Office of Administration
      and Resources Management
Director, Office of Transportation and Air Quality, Office of Air and Radiation
Agency Follow-Up Official (the CFO)
Agency Follow-Up Coordinator
Audit Follow-Up Coordinator, Office of Air and Radiation
Audit Follow-Up Coordinator, Office of Grants and Debarment, Office of Administration and
      Resources Management
Audit Follow-Up Coordinator, Region 9
Chief, Administrative Services, California Air Resources Board